First Responder Fire And Rescue Journal Notebook
Proud To Serve First Responder Journal Series
Gift Book For Firefighters And Fire Rescue Professionals
Paperback ISBN: 978-1-989733-41-7
Copyright Dunhill Clare Publishing 2020
All Rights Reserved. Cover Design by Sharon Purtill

www.ingramcontent.com/pod-product-compliance
Lightning Source LLC
Chambersburg PA
CBHW070150080526
44586CB00015B/1917